2276

Celebrating the Five Hundredth Anniversary of the Declaration of Independence July 4, 2276

Artell Smith

ISBN:

Paperback: 979-8-89669-534-9
Hardcover: 979-8-89669-533-2

Dedication

To my fourth great-grandfather, Thomas Downs, a soldier in the Revolutionary War (b. August 22, 1756, d. November 26, 1839). Originally from the County of Augusta, State of Virginia, but from six years old, a resident of Mecklenburg County (Rutherford) in the State of North Carolina. Thomas volunteered for service in the State Troop of North Carolina in 1776 in his twentieth year. The war record of Thomas Downs and a copy of his pension application documents are held in the National Archives in Washington, D.C. Thomas was the son of Zachariah Downs, who came to the United States of America in the early 1700s from Scotland seeking religious freedom.

Table of Contents

The Author

Philadelphia. April 1790.

The bedroom on the second floor of the house on Market Street was warm enough that the windows had been opened a hand's breadth, and the candle on the small writing table guttered whenever the spring air moved through the room. The old man propped against the pillows did not seem to feel the draft, though his daughter had earlier wrapped a second shawl about his shoulders and tucked a folded coverlet beneath his arms so he might hold the pen without his hand trembling against the page.

"You have written enough for this evening, Papa," Sally said, gathering the cup and the small dish of pap from the tray. "Dr. Jones said rest above all. Will you not put down the pen and try to sleep?"

He smiled at her, the loose-jowled smile she had known since girlhood and waved her off with two fingers. It was the same gesture, she thought, with which he had once waved off French ministers and English lords. "Soon enough, the pen shall avail me nothing, child," he said. "But the words I write shall outlast the breath that fails me."

She had already turned toward the door with the tray. She caught only the first half of the sentence and assumed he had agreed to set the pen down once she was gone. She would return in an hour with the laudanum and a little broth, and she had four other things to attend to before then: the fire downstairs, the message to her brother-in-law, the apothecary's boy at the kitchen door, and the linens. She closed the bedroom door quietly.

Ben listened until her footsteps reached the bottom of the stairs. Then he dipped the pen, drew a fresh half-sheet from the small stack at his elbow, and considered the empty paper for a long moment.

In the morning, he had finished a satirical piece of which he was modestly proud, and which he had set aside in the hollow beneath the floorboard near the foot of the bed. It was a small private joke he doubted anyone would ever find. Later in the afternoon, he put the finishing touches on his memoir.

The codicil to his will was complete and witnessed. His correspondence was answered, more or less. His worldly accounts were as settled as such accounts could be for a man who had lived in three countries and quarreled in all of them.

What remained, then, was the matter he had been turning over in his mind for some days, ever since he had finished the calculations on the compounded interest of the Boston and Philadelphia bequests and had seen, in the columns of pence becoming pounds and pounds becoming a fortune, how casually two centuries might be addressed by a careful hand. If two hundred years, he had thought, why not five? If a bequest of money, why not a bequest of something else?

He began to write, slowly, the pen moving in the slanted hand his physician had lately remarked upon.

One does not merely hope the world will be a better place for having lived in it. One does all possible things to bring about goodness and happiness for his friends, his family, and his fellow Citizens. And yet, as I near the end of my journey, which I do not regard as a burden but rather a mercy, I find I have hopes to share with those who shall rise in the dim mists of future time, perhaps even five hundred years hence, reckoned from that great Declaration of our Independence some fourteen years past.

Yes: that is the proper measure for what may be among the last offerings of my pen to the Reader. I shall take care that this short missive does not see the bright light of day until many years hence, after what I am certain will be many trials, tribulations, and toils for our young Nation and those who shall inhabit it.

Reader, I have lived long enough to learn that prediction is the comfort of fools and the ruin of philosophers. I make none. I offer instead ten hopes, modest, particular, and earned, for the Republic at her five-hundredth year. May she find them worth the small trouble of the reading.

The elder statesman set the pen down, briefly, to rest his hand. Through the open window came the sound of a carriage on the cobbles below, and somewhere farther off, a bell from one of the churches. He did not know which. He took up the pen again.

The Discovery

Old Washington. February 2276.

The cruiser made the run from New Washington in barely forty-five minutes, which gave Micah just enough time to decide, for the third or fourth time, that he should have talked Myra out of coming.

He did not say so. You did not talk Myra out of things. He had learned that much in two years at her side.

She sat across the little cabin with her face to the window, watching the flooded country slide by beneath them, enjoying herself enormously, which was somehow the part that worried him most.

"You worry too much, Micah. I'm old, not fragile." She did not turn from the glass. "I summited Kilimanjaro at thirty-one, and I'll tell you, the mountain gave me less trouble than this building ever has."

"I'm not worried about Kilimanjaro."

"I skipped Everest, you know. Too cold, too crowded, too expensive a place to die. A waste of a perfectly good view." Now she looked at him, bright with it. "You'd have hated it. All those people queuing up to suffer."

Micah had the building's last structural survey open on his handheld and was reading it for the third time. "I'm worried about a four-hundred-year-old cornice deciding that today is the day. If something comes down on the most decorated archivist in the service, on her last authorized expedition, I'm the one who files the report. I'd rather not write that sentence."

"There it is." She turned from the window, delighted. "He isn't worried about me. He's worried about the paperwork."

"I'm worried about both. In an order I'd rather not specify."

That got a laugh out of her, loud and unembarrassed, the laugh of a woman who had never once worried about the volume of it.

"I'll be sorry to see you go," he added, still not looking up from the survey. "I've also read the promotion criteria."

"Of course you have." It seemed to delight her rather than sting. "You and your criteria. I never read mine, you know. Fifty years, not once. I read maps instead. Footnotes. The backs of closets nobody else would crawl into. I'd have done the whole of it for free, and some years I very nearly did."

She went back to the window, content, and let him work.

Below them, the old country opened out, flooded and gray-green, the Potomac gone broad and formless, the tidal swamp it had been before the city creeping back over everything three centuries of engineers had fought to keep dry.

The Troubles had begun then, a hundred years and more ago, the long, bad stretch when the seas rose, the mountains woke, and the country very nearly came apart over who would drink what was left. Micah had been raised on the story, as every child of the Republic was. Myra had been raised nearer to it and did not talk about that part.

New Washington, where they were both born, sits safe and high in the Tennessee hill country, far from any coast, which is why it was built there. Old Washington: the Republic did not abandon so much as defend.

When the capital moved, the government would not give its monuments to the water, so it ringed the old core in levees, sea walls, and a great thirsty array of pumps, held the river off, and carried everything that mattered out at its own pace behind that wall.

The wall was always borrowed time. It cost a fortune to keep, and lately it had begun to lose, a dead pump here, a weeping seal there, until at last the engineers said the thing no one wanted said: they could no longer promise the core would stay dry, or that a party sent into it would come back out. So today was the last day. After this, Old Washington went back to the Potomac.

The last few trips into the diked city had brought home almost nothing. A Supreme Court justice's worn leather chair, one of the best of them. Micah did not expect today to beat a chair, had said so, and had been cheerfully ignored.

One thing behind the failing walls still kept its appointment with the spring. The cherry trees the Empire of Japan sent in 1912, as a token between friends, bloomed yet, every year, over ground that would not be dry much longer. Micah had heard you could smell them before you saw them.

The cruiser came in low over the Mall, the old white needle of the Monument standing up out of the diked city the way it always had and set down near the steps of the National Archives. Two escorts went out ahead, one of them a medic, just in case. The pilots stayed with the ship. The party had two hours.

They went up the long front stair, Micah a half-step behind her, watching her feet more than his own. The marble was cracked and weather-stained, but it mostly held.

Near the top, one slab did not. Her toe caught the lifted edge. She pitched forward, and Micah had her by the elbow before the fall was anything more than an idea.

She caught her balance, breathed out, and looked down at the slab that had nearly had her, almost admiring it. "Thank you, Micah." Then, to the step: "Good try." Myra laughed and went on up the last few steps, and Micah let himself breathe.

At the portico, the great bronze doors stood shut, as they had for decades, each nearly forty feet of metal that no longer cared to move. Service hatches had been cut into the marble on either side for parties like theirs. The lead escort keyed the right-hand hatch, and they went in.

Soon, the doors would be torn from their hinges and moved to New Washington, where they would stand as a statement about the importance of preserving the history of the United States and making it available to all citizens.

The air changed at the threshold. Cooler, drier, with the held stillness of a place being kept rather than lived in.

The dome was whole, though a long brown stain ran down from the eastern oculus and across the bare patch where one of the great murals had hung. Both paintings were gone now, lifted years ago to the museum in New Washington. Where they had been, pale clean rectangles remained, a shade brighter than the wall around them, and the small brass plaques that said what a visitor had once been able to see.

The cases that had held the Charters of Freedom stood empty on their dais, glass clouded but unbroken. The documents were long gone. The cases had been left because no one could quite bring themselves to take them.

Myra stopped at the center case and laid her hand flat on the rim of it, where the Declaration had lain once, and stood a moment as though she could still feel the warmth of it.

"You want to know what we're really here for."

"I've wanted to know since we left New Washington. You wouldn't tell me."

"I wanted you fresh." She grinned.

"Okay, prepare yourself," whispered Myra, with surprising reverence. "Here it is."

"At the end of his life, Ben Franklin left his papers in a glorious mess. Couldn't keep up with himself. Too many irons, too many countries. But he was careful, fussy even, about two things: his will and the money he left for Boston and Philadelphia, to sit and grow for two hundred years and then be spent."

She tapped the glass. "Now, Trustees are not historians, Micah. They look only at what they need to count the money, and at nothing else. So a thing could ride alongside that money for, well centuries, filed right next to it, and nobody with any reason to care would ever open it."

"And you think there's a thing."

"I think it would be exactly like him. Hide the good stuff inside the dull stuff and let the dull stuff stand guard." She was lit up now, the way Micah had only ever seen her when she had a genuinely good problem.

"An Easter egg, the old phrase was. Something tucked inside something else. I don't know what it is. I've never known what it is. I've wanted to know for thirty years, and today is the only day I will ever get to find out. So." She straightened. "Let's go and dig."

Micah checked the time. "We've got an hour and forty minutes of digging before the building decides our schedule for us."

"Then stop talking and follow me."

She led them down. The stair narrowed past the second flight, marble giving way to plain concrete, and a salvage crew's old emergency lights buzzed in the walls, dark in patches, so the lead escort lit a lamp. Myra went down past doors Micah would have stopped at without a second look.

"Where are we going?"

"Maintenance closet. Fourth basement, off a side hall." She said it the way another woman might have said the Crown Jewels.

"A maintenance closet."

"It isn't in any inventory as anything else, which is rather the point. Last anyone looked, it had the floor's cleaning supplies stacked in front of it."

"And you know about it how?"

"A footnote." She was nearly humming. "Second-to-last page of a dissertation. The man who wrote it came to me late in life to finish the thing, an old man studying what he had worked on as a young one. His whole subject was the loss of this building."

She turned into a side hall that matched none of Micah's maps. The escorts split to cover both ends of it.

"When the capital moved, you understand, they did not mean to lose this place, not at first," she said. "They walled the old core, pumped it dry, and carried everything that mattered out behind the wall, in order, over the years. But the wall was borrowed time, and everyone knew it, and toward the end, the work got frantic. The crews took what was on the lists and left the rest where it stood, for the water whenever the wall finally gave."

"And the boxes weren't on a list."

"The boxes weren't on anything. Old paper from a trust that had closed its books centuries before, shipped here and forgotten in a sub-basement. Nobody's idea of treasure." She kept walking.

"But he was young then, an intern barely, on one of those last crews, sent down to the bottom of the building to sort what was worth carrying out from what would be left behind for good. He found these. He couldn't take them; there was no room and no reason. But he couldn't quite write them off either. So he pushed them somewhere out of the way, where they'd keep, and told himself someone would come back for them before the end."

"And no one did."

"No one did. The wall held longer than anyone promised, and memory is short, and a few boxes of trustee correspondence are easy to forget when you have a whole city to grieve. But decades on, when he sat down to write the long sad story of how this place was lost, he set it down. Second-to-last page, a footnote nobody but me would ever read twice. Where he had left a handful of old Franklin trust papers on his last day down here, and whether, he wondered in print, anyone would reach them before the water did."

"And in thirty years you never came to look."

"In thirty years, I never got the chance. The Charters came first, always. Then the court papers, of which there are many. Then the wall got too uncertain to risk a party on a hunch."

She stopped at a narrow door with no plaque. "And then they told me this was the last trip and let me choose where we'd spend it. So." She laid her hand on the lever and left it there for a moment. "He's gone now, my old student. He never knew whether what he did that day mattered, or whether he only kept the water off them for a little while for nothing. I told him once that a footnote could outlive a man. I would dearly love to be right."

She tried the handle. It opened.

The room was a maintenance closet, eight feet wide and twice as deep. A utility cart held bottles of cleanser dried to crust, a stepladder leaned beside it, and a shelf of paper towels had gone brittle in their wrap. Three feet from the back wall stood a folding partition, the kind that once divided an office floor. The air was dry, cool, and dead still, a hundred and fifty years of pumps, levees, and pure stubbornness still holding the river off the bones of the building, though not, the engineers said, for very much longer.

"The back of it, behind the supplies." She read the closet against the footnote. "This is the place."

She set down her case and pushed the partition aside.

More of the same waited behind it. Cartons of tissue stacked to the shoulder, drums of cleaning concentrate sealed and never opened, far more than any building would keep for ordinary use.

"Hoarded." She rested a hand on a drum. "The shortage years. Somebody stocked this floor against a hard winter that was already on its way and never got to use a drop of it."

Micah turned his lamp slowly over the shelves. "This is a supply closet, Myra. There are no boxes. We have ninety minutes, and we are spending them on cleaning concentrate."

"We're spending them on the hunt."

"We're spending them on the finest stockpile of toilet paper in Old Washington." He did not bother to hide it. "I'd like to leave while the building's still willing to let us."

"Patience." She was already moving down the shelves, unhurried, reading the cartons with her fingertips. "The best part of any hunt is the minute just before."

She stopped at the far end, where a short stack of cartons sat apart from the cleaning stores, read the side of the top one, and went still in the way Micah had learned to watch for.

"These are paper." She ran a thumb along the edge. "Twenty-four-pound stock. Brilliant white. Georgia Pacific. They stopped making this before the Troubles. You could not buy it now at any price. The mills are gone."

"Is that it? The find is old paper?"

"It's a find. It isn't *the* find. But take a couple of reams. For science. Humor me."

She had already turned back toward the partition as she said it, half resigned, Micah thought, to the long climb out with nothing better to show for thirty years than two reams of dead stationery.

He lifted the carton lid and drew out the reams. As he turned with them, his shoulder caught a folding table leaning flat against the wall. The table shifted. Where it had stood, his lamp found a low gleam, a flat reflection where there should only have been a wall.

"Myra." He set the reams down and eased the table the rest of the way clear. "There's something here."

She came back.

Behind the table was a small door, no higher than his chest, the kind of hatch a building keeps for the parts of itself nobody is meant to see. Its handle was a plain steel lever; dull except for the one edge the lamp had caught.

For a moment, neither of them touched it. Then Myra worked the lever, half expecting nothing, and it opened.

The space beyond was barely a space at all, a pocket in the building's bones, dark and lightless and always had been. On the floor, stacked two deep and three across, sat six boxes, their cardboard furred soft at the corners, with the look of things moved once and forgotten on purpose.

Myra crouched at the low opening and did not reach in. She only looked.

"It can't be this ordinary," she said, almost to herself. "A closet behind a closet behind a closet, three doors deep, in a building this size, in a city the sea is owed." She shook her head. "He hid the thing inside the dullest letter he had. And then the years went by and hid the letter inside the dullest room they could find. I think he'd have liked that."

Then the wonder folded away, and the professional came back, and the work began. "Lamp close. Help me get them out where I can see, and gently, the corners are gone."

Micah, who had spent the whole descent counting minutes, found that he had stopped counting them. He brought the lamp close.

One at a time, they carried the boxes out into the larger room. The labels were in the cramped hand of an early nineteenth-century clerk, with trustee correspondence, City of Philadelphia, a stretch of years for each. Myra read them as they came, and at the fifth she stopped.

"This one." The yellowed label read: *Trustee Correspondence, City of Philadelphia, 1831 to 1837. Miscellaneous.*

"Why that one?"

"Because miscellaneous is where everything important hides. Two hundred years ago, the bequest ran out of Philadelphia. Whatever Franklin's executors filed with the original papers rode along with it, year after year, until some clerk in the 1830s swept everything older than forty years into one box under one word that meant he would never have to think about it again."

She knelt, with Micah's arm to steady her, and lifted the lid. "Help me." Inside lay two inches of folded paper. City correspondence. Ledgers. Copies of letters about the running of the trust.

"Mundane," she said softly. "Beautifully mundane."

She lifted the top leaves aside, then the next, then the next, each one handled with the same care and set down with the same small breath of not yet. Micah watched her hands. They shook, yet they were exact.

A quarter hour went by. She was two-thirds down.

"There may be nothing." She did not look up. "I have believed otherwise for thirty years. There may be nothing at all."

"Keep going."

She kept going. The paper aged under her hands, the ink browning, the rag of it coarsening. Minutes from the 1820s. A ledger leaf from 1812. A receipt for paper stock, 1801.

Then a folded letter, sealed once, broken once, and never folded back with any care, labeled in a clerk's hand as a note to a printer about engraving the trust's stationery. Housekeeping. Nothing. She nearly set it by unopened.

She opened it. There was something inside.

Where the old fold had eased, she could see the edge of a second sheet tucked behind the first, a different color, a different size, set where a careless eye would slide right past it.

She was quiet for a long moment. "I wonder," she said, "what we have here."

Myra drew the outer paper apart. Against the back of it lay a small bundle, wrapped in unbleached linen, tied with twine gone nearly the color of the cloth. The knot had not been undone in a very long time. She lifted it free and set it on her knees. She did not open it. She looked at the twine.

"Myra?"

"A moment, Micah. I have imagined this for thirty years. I would like the moment."

He waited.

She finally undid the knot. The linen fell open. Inside lay a folded half-sheet of laid paper, yellowed but whole, a second sheet folded round it as a wrapper. She lifted the wrapper away.

"Bring the lamp closer." He brought it closer.

The hand on the inner page was one she had seen three hundred times in her life. In the manuscripts at Penn. In the letters in Boston. In a half-dozen private collections. Slanted further than usual, as if written by someone whose physician had recently been remarking on the slant.

She read the first line. She read it again.

"Oh," she said.

That was all.

She raised her eyes to Micah's and held them, saying nothing about what she was looking at.

And Micah, who had come on this expedition because two years at Myra's side was the surest road to Myra's desk, who had spent the whole day on the clock and the survey and the line this trip would make on his record, found that none of it was anywhere in his head.

He had stopped thinking about the time. He had stopped thinking about the report. He was watching an old woman's hands shake over a sheet of paper, and for the first time, he thought he understood why a person might give thirty years to a footnote.

"What is it?" he asked. "Myra. What is it?"

"I think we call this in," she said at last. "Before we touch anything more."

He waited for more.

"You'll see it when we have it in proper conditions. It deserves that. So do you."

He nodded.

Twenty feet back, at the corner of the hall, the medic watched the two of them with the alert and unreadable face of a man trained to know the moment his ordinary day has quietly stopped being ordinary.

The Vote

Old San Juan. March 2276.

The classroom sits in the old wing of Escuela Superior Luis Muñoz Marín, two blocks up from the cathedral and four down from the harbor. The wing has stood a hundred and fifty years and been gutted and rebuilt four times. It keeps itself now at an even twenty degrees and opens its louvers to the trade wind only because the trade wind is pleasant, never because it needs the help.

Above the place where the room's voice gathers, an old, printed banner reads *We the People*. The three words have been the school's motto longer than anyone can say.

There is no front to the room. There has not been a front to a room like this in a hundred years.

The students lay, sat, and sprawled across the shared field, the soft globe of light the room projected at its center, paired to each student's personal interface. Through it, any of them could set a question, hang a citation, or lay a counter in text.

Diego had the broad window ledge, the harbor bright behind him. Ana Lucía lay full length on the floor, her hands laced behind her head, her eyes on the field from below, which was not laziness but simply where she did her best thinking, as everyone had long since stopped pointing out. Two more sat cross-legged on the warm tile. Sofía always stood.

Their emissaries were up and working. Each student kept one, a matte cube three centimeters to a side with a single dark lens for an eye. When not wanted, the cubes idled in the building's reserved band overhead.

All morning, the students had been sending them off on errands and questions that threaded the school unseen, so the argument in this room could run alongside a dozen quiet, private missions elsewhere in the school. A student could bring one down with a thought, and it would hover at a shoulder, attentive, then lift away.

Señora Profesora Iris Calderón was not at the edge of the room, because she was never anywhere for long. She prowled it. Late fifties, gray coming in at the temples, dressed in the loose linen the climate had made universal, and entirely unable to watch a good argument from a chair.

Behind her, on the wall where she kept her few private things, hung two doctoral hoods. The Sorbonne, where she had earned her Ed.D. in pedagogy. The University of Puerto Rico, where she had earned her Ph.D. in constitutional history. Both had been made possible by the Federal Teaching Corps

Scholarship she won at
once stopped being please

The doctorates could
research chair and a long
dead men and women. Sh
life and chosen a room
arguing at the top of thei
been sorry.

The students wore
holograms that ride a few
projected by each student'
the right shoulder, a shape
them stood in the talk, p
question pending, no urge
the room reads feeling rat
nothing yet. The left side i
It rarely stays that way.

Carolina reached into
question, turning slowly in
angle. *Cuba. Sexagésimo E*

Calderón let it turn a m
above it, in English, becau
two languages stacked toge

Are we ready?

The first glyph to brigh
green triangle. Triangle fo
would like to begin."

"Caro

"*Seño*

"You

Caroli
sentence,
Then she
she want
wanted c

"The
floor of t
committe
in 2238 a
substanti

She l
the whole
framewo
dispute.
Passage i

Unde
when wo
in the air,
room hea
a quiet c

Carol

"Cub
more fu
states m

through every door we left open, every year since 2236."

Continuing, Carolina said, "A Republic that will not grow by consent has ceased to be a Republic. The framers of the 28th Amendment wrote Section 1 with this day in mind. I will vote yes."

She drew her hand out of the field and sat. The pleased look stayed on her face because she had practiced that opening, and it had come out exactly as she had hoped.

Luz played the three notes back at her, slower, teasing. Diego bit back a smile. Carolina rolled her eyes at both of them, and for a second, the orator was gone, and there was only a girl of seventeen who had nailed her lines and knew it.

Then a green circle came up over the window ledge. Circle for a comment. "Diego, *por favor*," said Calderón.

He did not stand at first. He looked at his hands for a while. Then he got up, and the harbor was bright behind him through the louvers, and whatever had been teasing in his face a moment before was gone out of it.

"My great-grandmother was Lucía Esquilín Romeu. She captained the Esperanza out of Mayagüez. In April of 2227, three Cuban patrol boats boarded her in the Mona Passage. There was an exchange. Lucía came home. Three of her crew did not."

He let that stand in the air without sending a word of it to the field.

"She kept the boat another twenty-eight years, until her hands could no longer hold the wheel. I never met her. I hear that she never spoke about that day, not even to my father," Diego said.

"I am not against Cuba. Understand that. I am against the timing." He looked around at them. "Wait another ten years, or so. Wait until the last people who were on that water are gone, so that we are not asking anyone to applaud the thing who watched it happen. Then yes. Gladly, then. Just not now."

He sat back down on the ledge. Two of the left shoulders in the room had warmed to amber. Calderón let the silence do its work. She was very good at not filling silence, which, for a woman built the way she was built, had taken years of practice.

Above Ana Lucía, a soft blue circle appeared, a comment, no urgency. She had lain through it all, Carolina's case and Diego's grief, watching the field turn over her head.

"Ana Lucía," Calderón said, gently.

She did not get up. She spoke from the floor, eyes on the ceiling, which the room knew by now was how she said the things she had carried the furthest.

The Vote

Old San Juan. March 2276.

The classroom sits in the old wing of Escuela Superior Luis Muñoz Marín, two blocks up from the cathedral and four down from the harbor. The wing has stood a hundred and fifty years and been gutted and rebuilt four times. It keeps itself now at an even twenty degrees and opens its louvers to the trade wind only because the trade wind is pleasant, never because it needs the help.

Above the place where the room's voice gathers, an old, printed banner reads *We the People*. The three words have been the school's motto longer than anyone can say.

There is no front to the room. There has not been a front to a room like this in a hundred years.

The students lay, sat, and sprawled across the shared field, the soft globe of light the room projected at its center, paired to each student's personal interface. Through it, any of them could set a question, hang a citation, or lay a counter in text.

Diego had the broad window ledge, the harbor bright behind him. Ana Lucía lay full length on the floor, her hands laced behind her head, her eyes on the field from below, which was not laziness but simply where she did her best thinking, as everyone had long since stopped pointing out. Two more sat cross-legged on the warm tile. Sofía always stood.

Their emissaries were up and working. Each student kept one, a matte cube three centimeters to a side with a single dark lens for an eye. When not wanted, the cubes idled in the building's reserved band overhead.

All morning, the students had been sending them off on errands and questions that threaded the school unseen, so the argument in this room could run alongside a dozen quiet, private missions elsewhere in the school. A student could bring one down with a thought, and it would hover at a shoulder, attentive, then lift away.

Señora Profesora Iris Calderón was not at the edge of the room, because she was never anywhere for long. She prowled it. Late fifties, gray coming in at the temples, dressed in the loose linen the climate had made universal, and entirely unable to watch a good argument from a chair.

Behind her, on the wall where she kept her few private things, hung two doctoral hoods. The Sorbonne, where she had earned her Ed.D. in pedagogy. The University of Puerto Rico, where she had earned her Ph.D. in constitutional history. Both had been made possible by the Federal Teaching Corps

Scholarship she won at twenty-three and had never once stopped being pleased about.

The doctorates could have bought her a quiet research chair and a long career among the papers of dead men and women. She had taken one look at that life and chosen a room full of seventeen-year-olds arguing at the top of their lungs, and she had never been sorry.

The students wore their glyphs, the small holograms that ride a few inches above each shoulder, projected by each student's personal interface. Above the right shoulder, a shape that showed where each of them stood in the talk, pale and resting in blue: no question pending, no urgency. Above the left, where the room reads feeling rather than thought, there was nothing yet. The left side is always dim on the way in. It rarely stays that way.

Carolina reached into the field and set the day's question, turning slowly in the air, readable from any angle. *Cuba. Sexagésimo Estado de la Unión. Sí o No.*

Calderón let it turn a moment, then hung one line above it, in English, because she liked the look of the two languages stacked together.

Are we ready?

The first glyph to brighten was Carolina's, a small green triangle. Triangle for a question; green for "I would like to begin."

"Carolina."

"*Señora Profesora*, may I open?"

"You may. Take us in."

Carolina did not speak at first. She thought the first sentence, and it arrived in the field in her own hand. Then she said the rest aloud, because there were things she wanted hanging in the air as text and things she wanted carried in her own voice.

"The proposal for Cuba's statehood has sat on the floor of the Senate for two years. It has cleared every committee twice. Cuba's constitutional amendments, in 2238 and again in 2261, brought their charter into substantive alignment with ours."

She loved the phrase *substantive alignment*, and the whole room could hear that she loved it. "The trade frameworks have run for decades without a single dispute. No hostile vessel has entered the Mona Passage in forty-nine years."

Under her last clause, Luz, who argued in sound when words felt too blunt, let three rising notes hang in the air, a small, audible question mark that the whole room heard and understood. Tomás was already laying a quiet counter-citation in text along the lower edge.

Carolina ignored them both.

"Cuba is governed by a constitutional democracy more functional than half our own thirteen original states managed in 1789. Cuba has asked to join us

"My great-uncle was on the *Esperanza* in 2227. He didn't come home. My family has spent fifty years grieving him, and I have spent eighteen of them hearing about it. He was twenty-three." Without anyone telling them to, the emissaries began to come down.

Three first, then a fourth, then the rest, dropping one after another out of the band overhead, each student calling their own down with a thought, until the little cubes hung in a slow gathering above Ana Lucía with their lenses turned toward her. This was the room's version of a hush.

"My grandmother told me he would have voted *yes*." Ana Lucía's voice did not change. "He used to tell her that people who fish the same water have no business shooting each other over it. Cuba has been saying that exact thing, in writing, in their own constitution, since that dark time."

She turned her head on the floor and looked across at Diego on his ledge. "I'll vote yes. With both hands. Saturday."

The students' left shoulders stayed amber a long moment after she finished. The cubes hung where they were a beat longer, then lifted, unhurried, back up to the band. Tomás's right shoulder lit with a green square. Square for clarification, which was Tomás's natural habitat.

"*Señora Profesora.* The majority of Congress needed?"

"Simple majority. Both chambers, plus the President's signature. Exactly what Article IV has always asked. The 28th Amendment never touched it."

"And the majority we need? The people?"

"Simple majority of the votes cast, over the participation floor. Half the electorate has to turn out, and of those, more than half have to say yes." She held up two fingers, delighted by it the way she was every year. "So. Congress is one gate. We are the other. Section 2 in the 28th. Both gates have to open, or nobody comes in."

Tomás had pulled the section into the field as he asked, the text turning between them. "And if one of them doesn't?"

"Then it doesn't pass," Carolina said from the floor, not bothering with her glyph. "Either gate. Belt and suspenders."

"Belt and suspenders," Calderón agreed, and then she came off the wall, because they had reached her favorite part and she was incapable of staying still.

"Oh, this is the part I love. Section 3 was one of the bloodiest fights in the whole ratification of the 28th. Months of it. People who agreed on everything else nearly walked out over this one thing. And do you know where they landed?" She turned a hand through the field, and the section turned with it. "Nowhere. They took power away from nobody. Congress decides exactly as Congress always has."

She looked around at them, lit up.

"The 28th was ratified by Congress and three-quarters of the State legislatures, the way amendments are. And what it did was widen the door, not raise the bar. They put the people in the room next to Congress and required the two to agree. So why on earth would anyone build it that way? Harder to pass, not easier to break. Somebody tell me why."

A few glyphs flickered and thought better of it. Then Daniela put up a soft blue circle and did not wait to be called, which Calderón allowed with a nod.

"Because if it's hard to get in," Daniela said slowly, building it as she went, "then everybody believes it once you are. Nobody can say afterward that it was a trick, or a bare majority on a bad day. It costs more, so it counts for more." She winced. "That sounded better in my head."

"It sounded perfect out loud," said Calderón. "What else?"

But Daniela had spent her courage and looked glad to hand it on with a question. "What about the new citizens, though. The Cubans. On the actual day."

A classmate answered before a glyph could even form, the words landing in the field a half-second behind the voice. "All of it. Day one. The whole Bill of Rights, the moment they're admitted. No tiers, no waiting period, no probation. That's what Section 4 was for. We did the other thing once, in the territory years, which didn't work well."

"Rights are floors," Calderón said, and for once she said it quietly, because some things land harder when spoken softly. "The whole Bill of Rights is a floor."

A small collective groan from the back because they had spent two months on the first ten amendments to the Constitution last semester and were not eager for an encore. Calderón did not look back at them.

"The people can raise the floor by referendum any time they like. They cannot lower it. Not by a majority. Not by every last vote in the Republic. Section 4 is the firewall, and it only burns in the one direction."

Carolina had gone quiet a while back. Her glyph was dark. She sat with her arms around her knees, watching the question turn, and you could see her chasing something around the inside of her own head.

She said it half to herself and half to no one. "I mean... we could always... if it really, truly didn't work out..." She trailed off.

It was Diego who answered her. Diego, of all of them, from the ledge. "No. Section 5 in the 28th." He was not unkind about it. "Once they're in, they're in. There's no undo. Section 5 isn't a loophole somebody forgot to close. It's the whole reason the vote means anything at all."

Carolina looked up at him and held his eyes a long moment.

The room had gone very quiet. Several left shoulders were amber. Nobody put a thing in the field. The question turned slowly in the middle of them all, still unanswered, and Carolina sat with the thought the room had just walked her into. One ordinary majority, taken once, on one ordinary Saturday, that no later majority could ever take back.

Calderón said nothing at all. The argument had made itself, which was the only kind she altogether trusted, and she would not have stepped on it for the world.

Then Sofía's right shoulder lit with a deep blue triangle. I have prepared remarks.

Sofía was the youngest of them, the last to come of age, and she would manage it by exactly one day. Friday was her birthday. The whole room knew it. Calderón knew it.

"Sofía."

She had been on her feet by the louvers the entire class, and she stayed on them. "I have been a citizen of this Republic since the day I was born. I have voted in classroom proceedings since I was five. I have argued this precise kind of question with these precise people for thirteen years, and I will tell you, some of you have not gotten any better at it."

That got a laugh, even out of the two amber shoulders.

"On Friday, I turn eighteen. On Saturday, I get to do it for real, by a single day." She steadied herself. "And I will say honestly that I do not yet know how I will vote. I am certain of only one thing, which is that the question is a great deal bigger than my opinion of it."

"*Mis abuelos* lost cousins in the Mona Passage. *Mis abuelos* also raised me to believe that fifty years is fifty years, and that a Republic that cannot forgive is not worth the trouble of defending." For just a second, she looked five again, on her first classroom glyph. "I think I will vote yes. But I am still listening."

She sat on the ledge beside Diego. The ten-minute warning arrived, a soft tone that the room let in the way it lets the wind in. Calderón gave them that time to confer in pairs.

The big field broke into a dozen small private ones as the students turned to one another, the cubes dropping out of the band and pairing off above their owners, and the room filled with low overlapping talk and the odd burst of laughter. Diego and Ana Lucía drifted together, sat comfortably on the floor, and what passed between them stayed between them.

Carolina crossed to Luz, and the two of them put their heads together over something that made Luz play a rude little four-note phrase, which Carolina swatted out of the air. Sofía went to Diego and put a hand on his shoulder, and his left glyph flared amber once and went dark again.

Tomás came over to where Calderón stood. His glyphs were off, and his emissary had folded itself onto his shoulder, quiet.

"*Señora Profesora.*"

"Tomás."

"I sent my Teaching Corps application in yesterday." Calderón stopped what she was doing.

"You did not!" She was grinning now. "Tomás. Which program? Tell me all of it."

"Constitutional history. Penn. I put Cantonese for the language requirement."

"You'll get in."

"My mother says the same thing."

"Your mother," said Calderón, "is a woman of excellent judgment."

He almost smiled. Then he hesitated, and the procedural confidence went out of him all at once, and he was only a kid asking the thing he had really come over to ask.

"*Señora Profesora.* Can I ask you something?"

"Always."

"Did you ever doubt? Doubt your, uh, qualifications? On your first one. The first real constitutional referendum you ever voted in."

She thought about it honestly. Behind him, the room was thinning, the small fields winking out one by one, the cubes climbing back to the band as their owners gathered their things.

"I did," she said. "Badly. And I voted anyway. That is what the day is for, Tomás. Doubt is not the thing that disqualifies you. Doubt is the dues. The morning it stops being hard is the morning you've stopped taking it seriously."

Tomás nodded. He thanked her, and he went, his emissary trailing up the air behind him. And then she had the room to herself.

The field at the center had gone clear, only the day's question still turning where Carolina had set it that morning. Calderón reached in and brought up her own civic credentials on the resting screen behind the field, the ones she would use to vote on Saturday.

Cuba: Admission as the Sixtieth State of the Union. Yes or No.

She had not opened it yet. Two doctorates and a whole career spent teaching other people how the machine of consent was built, and she still was not quite ready. Calderón smiled at that and closed her field.

The *Señora Profesora* gathered her notes and a light shawl. The louvers stood open on the harbor, the trade wind coming through warm and going again, the room holding its even cool behind it. Saturday was four days off yet.

She had time. And the doubt, she thought as she went out, was sitting exactly where it ought to be.

The Tether

Sweetgrass Valley, Colorado. May 2276.

The frames were still running when Nora came up the terrace road at dawn, and the valley shimmered the way it had every morning of her life.

She stopped the cart at the crest and looked down. The terraces fell away in long, shallow steps, and over each one the air trembled, the way it does over a hot road in August. Amid the trembling, the grain stood waist-high and faintly lit, its amber pooling in the low places.

In three days, the valley would belong to the Tether.

Nora was forty-one and ran the farm. Her grandfather had taught her every part of it. Christian Whittier was past eighty now, his hands too unsteady for the fine work, but his judgment had not slipped a day.

The Whittier place was one of a hundred and nineteen farms in the valley. The people arguing about Sweetgrass down in the cities pictured one stubborn clan on an ancestral spread. It was never that. It was a hundred and nineteen families, and no two had come by the same road.

The Haugens came down from Minnesota five generations back, Lutherans who never forgave the altitude. The Trujillos carried their name up from the old Spanish towns to the south. There were the Begays and the Okonkwos. Two Latter-day Saint families ran the best orchard on the east terraces.

There was a Catholic chapel, a plain white meeting house that served as whatever anyone needed on a given Sunday, and a good third of the valley that kept no church at all and came to the potlucks anyway.

They had all come a hundred and seventy-six years ago, when the land office opened the high parks to anyone who would buy a hundred acres, lease the range around them, and grow grain where the maps said it could not be done. Most came for the farming and stayed for the neighbors, which was lucky, because the farming was only ever average. Sweetgrass fed people. It never made them rich.

And the children left. That was the quiet truth beneath two years of noise. They grew up in the shimmer, then went down to Denver, Colorado Springs, Boulder, and did not come back. Nora's own two were in Denver now. Half the farmhouses held one old couple and a row of cold bedrooms.

Sweetgrass had been getting quieter on its own for a long time now. The Tether only put a date on it.

They had already moved the dead. Last month, the crews came and lifted the graves, gently, every family's, and carried them up to the new ground on the rim, the one piece the families would keep. The men were unhurried and wore no expression that asked anyone to feel a certain way. The Republic, when it took your valley, did not also ask to be thanked. There was a mercy in that Nora had not expected.

That same evening, those who were left gathered in the meeting house. It began civil but did not stay that way.

"We refuse." Dell Hauser was up before the chairs stopped scraping. "Let them come carry us out by the arms. We'll take it to the Supreme Court."

"We already took it to the Supreme Court, Dell," Christian said without getting up. "Eighteen months ago. We lost, eleven to nothing. The referendum was the last card we had, and we lost that too. There is no door behind this one."

The Pruitt boy's mother said they needed more time. You couldn't ask people to walk off almost two hundred years of work in one season.

"We've had two growing cycles," Christian said. "Two years of knowing. They gave us that on purpose, so no family would be put off its land in a panic. Time is the one thing they did not stint us."

"It's the money." Hauser again. "They're not paying what this ground is worth. Not close." A few people laughed, which was not kind but was also not wrong to do so.

"Dell," Marisol Trujillo called from the back. "They are giving us a range twice the size of what we leave. Good ground, proved ground, down on the western slope. My family ran the numbers four times because we didn't believe them either. It is not a robbery. It is the best land deal anyone in this valley will ever see, and you want to turn it down out of spite."

"It's an insult to the ones who came before us." This came quieter, from the back, and landed harder for it. "All of us have people in that ground. And we hand it over because a vote in Denver said so?"

The room went still. Christian let it stay still, because the man was not wrong about the grief, and the grief had earned the quiet.

"I've thought about the ones who came before us more than any of you," he said. "On account of being closer to joining them." A small laugh. "And I'll tell you what I know about them, because I knew a few, and you know only the stories. They were careful people. Practical. Their ancestors bought a valley the maps called worthless and made it work, and not one of them was ever a fool about a deal."

He looked at the man in the back.

"Put this offer in front of my grandmother, twice the land for ground the country needs, and we cannot keep, and she'd have signed before supper. The way you honor careful people is not to throw a tantrum in their name. It is to be careful, the way they were."

It was Damian's boy who finally said the true thing. He was nineteen, going down to Boulder in the fall like the rest, and he loved the valley the helpless way you love what you are already half gone from.

"We did everything right," he said. "We made the case. We went to court. We voted. And they took it anyway. How is that anything but theft with the paperwork filled in?"

"It's the system working. That's the hard part." Christian leaned forward on his stick. "We voted, and we lost. The day a man honors only the votes he wins is the day he stops being a citizen and turns into a customer."

"A customer can walk out of the store."

"He can. That's the whole difference. A customer takes his business elsewhere. A citizen has nowhere else to take it. There is no elsewhere. There's only here, and the people we share it with, and what each of us gives up so the sharing holds."

"We settled the question of walking out a long time ago, at a price you've read about, and I heard from people who lived it. We don't get to unsettle it now because the lot fell on us."

"Then why us?" the boy said. "Why here?"

"Because there's nowhere else. You've read the survey. They need ground that's high, flat, dead quiet underfoot, solid all the way down, with empty sky over it, in country no city will ever crowd. There are only a handful of such places on the whole continent, and all but one has a fault under it or a town on top. We're the ones that're left."

He shook his head, half amused.

"We didn't lose this valley because it failed us. We lost it because it's the one piece of ground in the country where the thing can stand, and our great-grandparents put wheat on it a hundred years before anyone knew that mattered."

He closed his weary eyes. Then, quieter.

"And we weren't the only ones asked. There are three more on far sides of the world. A valley in the Asian uplands. A headland on some cold northern sea. A high plain in the south. Three more towns of people no different from us, who moved their own dead last month, same as we did, so all four roads could be built. We aren't a valley that got picked on. We're one of four doors the whole of humankind agreed to open. Ours is the only one that's home."

"They're going to build the climb here." For the first time, something besides patience came into his voice. "Right here. The road up. People and cargo lifted off this planet without being burned into the sky, the way it was always meant to go and never could."

He lifted a hand toward the north.

"The four roads arc up off the curve of the world and meet over the top of it, over the pole, where no flag was ever planted and no country owns a foot of ice. There's a thing up there that holds its place on power I won't pretend to understand. A crown of light over the top of the world, with four bright threads climbing to it. Ours will be one. From there a person can go anywhere. The Moon. Mars. The Belt."

He looked around the room.

"This ground fed people for two hundred years. Now it's going to lift them, and not just our own, because the crown it reaches belongs to nobody. That is the only reason it can belong to everybody."

"It's the largest thing this valley will ever do, and we're the last people who'll ever farm it. Both of those are true in the same hour. A grown person can carry both."

Nobody answered him, because there was nothing left to answer.

On the last night, while the families went down to the Church for the final supper, Christian asked Nora to walk back up to the home terrace with him. The light was going. The grain held it longest, the amber deepening as the sky cooled, so that for a little while the valley seemed lit from below.

"Shut them down," he said.

She had known he would ask her, and not do it himself, and she understood why. He had powered the frames up ten thousand mornings. He did not want the last act of his hands here to be the one that ended it. He wanted hers. She was the one going on.

Nora did it the long way, terrace by terrace, the way you put a herd down for the night. There was no reason to be gentle except that gentleness was the only thing left to give.

The home terrace first. The shimmer thinned, wavered, was gone, and the grain settled a fraction of an inch as the full weight of the world came back into it for the last time. Then the second terrace. Then the third. The trembling left the air step by step down the valley, and the light went ordinary, and it was only grain in the dusk, green-gold and heavy and still.

When the last frame went dark, the valley was quiet in a way Nora had rarely heard. The frames had always made a sound so low you noticed it only when it stopped.

It stopped now.

Christian looked at it for a long while.

"They'll raise it right about there." He lifted his stick toward the center of the valley, toward the flat sound ground where the grain stood heaviest. "Where the wheat is best. Of course."

Something close to a smile crossed his face.

"This valley carried grain up out of nothing for a hundred and seventy-six years. Now it'll carry us the rest of the way. Same ground. Same trick. It just reaches higher the second time."

He turned and started down, slowly, his hand on her arm. Nora went with him. Neither of them looked back, because the valley was as it had always been, and they wanted to keep it that way in the one place it would go on existing now.

The crews would come in two days. They would not be there to watch.

But on a clear night, in years neither of them would live to see, their grandchildren would stand on the rim where the graves had been moved and find it overhead. A still point of light fixed above the top of the world, with four faint threads climbing to it off the curve of the Earth. One rose from where the wheat had been.

The other three, from places in other countries, including homes that had been somebody's whole world too, had given up the same thing for the same reason.

The crown they held up together belonged to no nation. It had cost four peoples, to hang one thing in the sky that none of them could ever call their own.

And from it, a person could go anywhere at all.

The Rock

Interamnia, the Asteroid Belt.
June 2276.

Interamnia was never the brightest gem in the Belt, but she did the hard work.

A carbon-and-silicate giant three hundred kilometers across, she gave up her riches slowly and without glamour. No gold, no platinum, none of the bright metals that made a person rich in a season.

What she held was duller and worth far more. Carbon, feedstock, the bulk stuff that fed furnaces and farms and built the bones of every settlement for a hundred million kilometers in any direction. A larder, not a vault, and the place everyone eventually had to come to.

American companies were the backbone of Interamnia operations. ExxonMobil, Arch, Peabody, Energy Fuels, the old carbon and energy names that had outlived the fuels they were born to extract, hauling across the sky the carbon they'd once torn out of the ground.

Fifty years they had shared the rock, and got on well enough, as rivals will when the work is good and the margins better. Old loyalties ran deep, and there was always some squabble for the United Space Agency to settle, a claim line, a water ration, a slight remembered from a boardroom long ago. But nobody was yet at anyone's throat.

Under them, between the shafts and the smelters, a town had grown. An American town under rock and vacuum, with a school, a chapel, and a square, the USA keeping the peace over the whole of it. Families had made the barren place a home. That was what was worth keeping. Not the carbon. The town.

Two levels down, in the great vaulted square they'd hollowed out of the heart of the rock, a seven-year-old George Washington was having trouble with his wig.

"It won't stay on, Mom."

"Hold still." Margie Delgotti pinned it again, kneeling on the stone in the blue dress and white cap she'd sewn herself. Betsy Ross for the third year running, and getting, she'd have told you, no better at it. "There. General Washington. Try not to lose your hair on the day of the parade."

Calvin turned to study himself in the dark glass of a shopfront. Behind him, the square was strung with carbon fiber bunting and half-painted floats, and a banner two hundred feet long that read FIVE HUNDRED YEARS and wasn't finished yet.

All of Interamnia's residents had been at it for weeks. In a few more, on the Fourth, every soul would come down here to march and sing and remember a piece of paper signed five centuries and five hundred million kilometers away, by men who could not have imagined this rock, these children, or the black outside.

Calvin had been born in Interamnia and had never once stood on any other ground. The square and the shafts and the dark past the walls were the whole of the world to him, and his father meant to keep it safe long enough for the boy to grow up and go see the rest.

For Colonel Delgotti was no officer passing through on a short tour. He had made his life here, a wife in hydroponics, a son in the third grade, a town that was his to hold. In an older story, on an older frontier, they would have called him the sheriff, and they would have been near enough right.

He watched his wife and his son from the edge of the square and then made himself stop. The message in his ear had come twenty minutes ago, and he hadn't been able to set it down.

Delgotti found Hollis in the USA office, a glowing track pad between her hands, wearing the look company security people get when they'd rather be wrong. Hollis wasn't USA; she'd been appointed by the companies to monitor employees, immigration, and communications.

"Colonist liner," Delgotti said, reading it.

"That's what she's squawking." Hollis didn't sit. "*Serene Leader*, out of the Kingdom. Two hundred aboard, everyone logged a colonist, every paper clean. We'd have waved her through and poured the moonshine."

"But."

"But we track everything that comes within a week of this rock. That is how four proud companies have shared it for five decades without a civil war." A thin smile.

"And the *Serene Leader* is not what she says she is. Two hundred colonists, and the manifest reads like a military barracks. Three men for every woman, almost no children." She touched the pad and a column of figures rose out of it. "Colonists bring children, Colonel. That is the whole point of being colonists."

"And then there's the cargo. A colony ship carries seed stock and hab frames, the dull little miracles that turn a rock into a home. This one is hauling four hundred tons of equipment nobody has bothered to itemize. Crates that match nothing in the catalog. We asked. They declined to say."

"Could be smuggling?" said the Colonel with a lifted brow.

"I'd love it to be. But smugglers run for the towns that look the other way. They don't point two hundred fighting men and four hundred tons of nothing-you-can-name straight at the one rock the whole Belt can't live without." She let it sit. "And it's the Kingdom's flag on the hull."

Delgotti looked at the slow, bright mark crawling in out of the dark, and thought, against his will, of a banner that wasn't finished and a boy who couldn't keep his wig on.

"They never signed the Accords," he said. "I can't fire on them for a bad feeling. Everyone gets to leave, that's the rule, and it is a good one, and the day Command starts shooting colony liners out of the black on suspicion is the day we turn into what we are out here to stop." He was quiet a moment. "How long until she's here?"

"Six days."

"Then she doesn't get here." Delgotti was already moving. "Get me a frigate. We'll meet her in the dark."

USA Peacekeeper VIII met the *Serene Leader* four days out, in the long, cold dark, where there was no town square underneath with happy, ordinary people, doing mundane things.

Captain Park hailed Interamnia first, not the frigate, as though the warship sliding up his flank were a formality he could talk past. His voice came through smooth and sure. "Interamnia colony, this is *Serene Leader,* two hundred souls inbound for settlement. Send landing coordinates for our shuttles. We carry colonists and supplies, and we have come a very long way."

Delgotti answered from the frigate. "*Serene Leader*, this is Colonel Delgotti, United Space Agency. You will have your coordinates, and your shuttles will be welcome on the surface the moment you let us aboard to look at your ship. All of it. Including the four hundred tons of cargo, your manifest cannot seem to name. Open the boxes, Captain, and Interamnia is yours to land on. Keep them shut, and it isn't. No dice."

The voice that came back had a temperature to it.

"You will not board this vessel." The smoothness was gone. "These are families, Colonel. You have no authority to search a peaceful migration, none, and you will send the coordinates now or answer to my government for turning women and children out into the dark."

"Your government didn't sign the Accords, Captain, so I'm not sure what answer I'd owe it." Delgotti kept it flat. "But I'll make the offer twice, because it's a long way home and I'm a patient man. Let us look, and you land. That's the whole of it."

Peacekeeper VIII slid closer, near enough now that the liner would feel her without using sensitive sensor arrays. "I'm putting a boarding party across, Captain Park. You can open your locks and make it easy, or you can make it hard. There's no version where you reach Interamnia without my people seeing inside that hold."

The *Serene Leader's* answer was a plasma cannon.

A beam went out across *Peacekeeper VIII's* bow, close, a hard bright line in the black that any captain in the system would read the same way. We are not afraid of you. We mean to come through. It was the sort of thing the Kingdom drilled into its officers, and the sort of thing no one had done to a USA frigate in a long while, because everyone raised out here knew what a frigate was for.

Peacekeeper VIII did not fire back. Frigates didn't. Not first, not at a hull full of people, not ever if there was another way. What she did instead was speak once, in the only language that ends an argument without ending a life.

The pulse, unseen, crossed the gap in no time at all. Aboard the *Serene Leader*, every screen went black, every drive went cold, every gun went deaf in its mount. The ship hung there blind and drifting, with one system left running, the one the Agency always left running on purpose. The air. Two hundred people continued to breathe.

Delgotti crossed with the boarding party himself. He didn't have to. Younger men and women were better suited to airlock breaches than he. But it was *his* rock at the far end of this, and *his* boy in a wig two levels below it, and he'd never learned to send people to do what he wouldn't do himself.

USA synthetics flew the approach, scanned the hull, and found the soft places. The men went through the locks. A machine could open a door. Only a person could decide what to do with the man on the other side of it.

They met resistance in the corridors and put it down without killing anyone, which took longer and cost more, and was the whole difference between them and the men they were arresting. By the time Delgotti reached the bridge, his people had it secured, and Captain Park stood in the middle of his darkened ship with two USA Marines on his arms and murder in his face.

"You have no right. This is an act of war."

"You fired on a USA frigate, Captain." Delgotti looked past him to the men along the bulkhead, the haircuts and the hands, not a child among them. "I'd think hard about which of us you accuse of starting something."

The ship's hold told the rest. They opened the boxes Park would not name and found what a manifest of seed stock and habitat frames is meant to hide. No seed. No frames. Breaching charges, hard armor, and the long ugly tools of men who come to take a place and hold it against the people already living there.

A sergeant put a few questions to the youngest of the soldiers, and they answered the way frightened men do when the operation has already failed, and the lie is strewn across the deck around them. It had never been a settlement. It had been a seizure. The larder of the Belt, taken at gunpoint and kept.

"Here is what happens now, Captain Park," said Delgotti. "When your systems come back on, and they will, you turn this ship around and find somewhere else to point it. Not here. Interamnia is taken, Captain, by people who got here first and built something and mean to keep it, and you are in violation of the Accords from your drive bell to your nose."

"My government never signed your Accords."

"I don't care that your government hasn't signed the Accords." He said it without heat. "That's not why you can't have this place. You can't have it because it's somebody's home, and you came to take it, and out here that is the one thing none of us will abide, paper or no paper. You're not welcome. Maybe some other rock will take you in."

More thoughtfully, Delgotti continued, "I doubt it. But that's not mine to solve. Go find your own rock, Captain, and plant your own town on it with your own hands, the way these people did. Or go home. Those are the roads. There's no third option."

Then he turned to his sergeant. "Strip the hold. Every weapon, every charge, every round. And dismantle the main ship's weapons array. All of it. When the lights come back on this ship, I want her able to fly and breathe and steer, and not one thing more."

Park found his voice. "You cannot disarm a sovereign vessel."

"I can't turn an armed one loose to go hunting a softer rock than mine, either." The Colonel didn't bother to look at him. "You can sail wherever you please, Captain. I told you that, and I meant it."

"But you'll go with empty hands. Whatever you had planned for the next colony that wasn't ready for you, you'll have to manage some other way, because you won't be doing it down the barrel of these." He nodded at the open crates. "That's the one thing I can hand the rest of the Belt before I hand you back to it."

And then Delgotti did the thing Park had no box to put it in, the thing the Hermit Kingdom would never quite manage to explain to its schoolchildren.

He let them go. Two hundred of them, the soldiers and the captain who had fired on his ship, sent off alive and toothless into the dark to try their luck somewhere that wasn't his.

Delgotti stood on the *Serene Leader's* dead bridge with his family conjured up from memory in front of him, wanting in his heart to make sure these men never threatened *anyone* again, and yet he chose the other thing. Not because he was soft. Because the alternative was to become the very thing he had crossed all that black to stop, and a man who does that has won nothing worth keeping.

Peacekeeper VIII arrived home two days later. When Delgotti rode the lift down into the noise and the half-hung bunting, his son ran at him across the square in a wig that still would not stay on, and Margie looked up from the unfinished flag, and for a moment he stood in the middle of the very thing he had gone out into the dark to keep, and it was enough.

The Synthetic

Albany, New York. June 2276.

Loudonville was about eight kilometers northwest of the city, behind old trees, in the kind of comfortable quiet where the houses stand well back from the road and the road itself seems to apologize for passing through.

Gerald left the Olin house a little after four. The fast line out of the village would put him downtown in under five minutes, and he liked the fast line. He liked most things, if he was honest, and had decided long ago that he might as well be honest, at least with himself.

He did not have much time to think on the train, so he did what he always did: people watch.

A family sat across the car from him with a mother, father, and a boy of about nine, with the bright, loose, happy look of people at the start of a trip. A holiday, Gerald decided. He was rarely wrong about such things and took no pride in being right; a hundred and fifty years of watching people will teach anyone to read a family across a train car.

The boy's backpack was covered in stickers. The newest one, still bright where the others had gone soft with washing, read National Baseball Hall of Fame and Museum, Cooperstown, New York.

The boy kept turning to the window, as if Albany might come sooner if he watched hard enough, while his parents talked over his head about what they would see.

"We have to see the Capitol building," the father said. "They built it by hand. Took them thirty years." The boy did not care about the Capitol. The boy wanted to know if there would be time for the river.

Gerald found himself smiling and saw no reason to stop.

Seventy-six years is a long time to keep a single afternoon, but Gerald had kept longer, and one day came back to him now, whole. He had taken young Henry Olin to Cooperstown the summer the boy turned ten, just the two of them, which had been Henry's idea and not the family's. They stood a long while before the old gloves, the old bats, the photographs of men dead a hundred years, or two hundred years, and Henry had gone quiet in the way a child does when he first meets something larger than himself.

Henry ate too much that day. He filled himself with three stadium hot dogs and a paper boat of fries, was mildly sick on the ride home, and pronounced it the best day of his life, and meant it. He was an old man now, with grandchildren of his own, and he still brought up that afternoon during the Winter Holidays and still got one part of it wrong every single time, and Gerald had never once corrected him, because the way Henry remembered it was better than the way it had happened.

The train rose into the light at the downtown station. Albany has been the capital of New York for the better part of five hundred years, and it wears the fact easily, the way old places do. Gerald stepped off, walked the two kilometers to the Bureau without hurry, and arrived twenty minutes early, which was on purpose. He did not like to make anyone wait. Twenty minutes early, he had found across a great many years, was about the right width for courtesy.

The Citizenship Application Bureau kept a regional office in a Federal building above the Hudson, one of the plain ones that went up in a hurry during the rebuilding, after the Troubles, and was never loved by anyone. The intake hall on the ground floor stood nearly empty. Hardly anyone came in person anymore; the work had long since passed to soft-spoken holograms and applicant-counseling intelligences that could sit with someone in forty languages at any hour and did so kindly and well.

Gerald could have done the same. He could have sent a figure of light down to that hall and never left Loudonville at all, or the humans he considered to be his family.

He had chosen to come in person. He knew the appointment would be an odd one for whoever drew it, a flesh-and-blood examiner doing the work by hand at the end of a long day, and he did not wish to put one inch more distance between himself and that person than the day was already going to ask for. It seemed to him a poor way to request being treated as a person, to arrive as a trick of light.

So Gerald he had ridden up out of Loudonville, walked the two kilometers, and now sat in a hard chair in the quiet hall with his hands folded, waiting to be called, and was, on the whole, hopeful.

***

The last appointment of the day was a name Hale did not know, beside a category that made no sense, and he assumed, the way you assume small things at the end of a long afternoon, that it would sort itself out once the applicant sat down.

He had been an examiner at the CAB for eleven years. In that time, he had made several thousand people citizens, and had learned that the whole of the job, for all its forms, came down to one quiet act repeated across a desk. A person said, more or less, I would like to belong here. And Hale, on behalf of the country, said yes.

Hale had not expected to love the work, but he did.

He woke the desk, and the appointment opened in the air above it in pale light.

The applicant who came in was tall, unhurried, and, Hale saw at once, beautiful. It was the beauty of the better synthetics; a face built a degree more even than nature troubles to make one handsome when handsome was the instruction. He had dressed with care, in a dark coat a little out of fashion, the way someone dresses when the day matters to him.

And beneath the left eye, small and pale and impossible to miss once you knew to look for it, sat the short row of figures the law required every synthetic person to carry and forbade any of them to remove, alter, or hide.

A human being might be a great many things in that chair. A human being could not be that. So Hale knew what he was dealing with before his visitor had crossed the room.

"You're here for the Olin family," Hale offered. He was reading the line that had not made sense. Synthetics ran errands; it was a fair guess.

"No, sir. I am here for myself." His visitor sat, folded his hands, and offered a small, easy smile. "Which is the unusual part, I know, so I won't pretend it isn't. You can see what I am. I'd only waste the time you've set aside for me by talking around it. My name is Gerald Olin, and I have come to apply for citizenship."

Hale sat back.

In eleven years, no synthetic had ever said that to him. The regulation that answered it rose in his throat at once, automatic. He did not say it. He told himself later that he had only been too tired to be short with anyone, and that was true, though it was not the whole of the truth.

"Mr. Olin," he said instead.

"Gerald, if it's easier. I answer to both. I have answered to Gerald for a hundred and fifty years." He said it lightly, the way a man tells an old joke on himself, with no bitterness in it that Hale could find.

"It was the first Mr. Olin's grandfather's name, the gentleman who had me made. I have his name, and his chair at the holiday table, and none of the rest of what he had. But Gerald is fine."

Hale reached into the light and drew the form up between them. It had been built by people who never once imagined a day like this one, and every field on it showed as much.

"You understand there's no procedure for this."

"I understand there is no procedure for anything until the first time someone does it. I am aware I may be asking you to be the first time. I am sorry for that. It is a great deal to put on a Tuesday."

Hale nearly laughed. He did not, quite, and started on the form instead, because his hands wanted something to do and because some part of him had already decided to take it all down, whatever came of it.

"Full name."

"Gerald Olin."

"Date of birth."

"Date of manufacture, if we are being strict. The twelfth of February 2126."

"Place of birth."

"A facility in Paterson, New Jersey. You may put down Paterson. It is accurate, and a good deal nicer to say."

The form wanted a physical description. Hale glanced up to take one, only to find he did not need to.

"My eyes are this color because the first Mrs. Olin chose it from a chart," Gerald said, catching the glance and not minding it. "She had good taste. It is the one feature of mine that anyone has ever opened a catalog to select. The rest of me they left to the standard model."

"Occupation."

"That one will take a moment." Gerald considered it, not because he lacked an answer but because he meant to give a fair one. "I keep the Olin households. Three of them now, the principal one up in Loudonville. I tutor the children. I have taught three generations of Olins their letters and figures, and two of them hold doctorates today, which I mention only because it is the part I am proudest of. I keep the family's accounts, calendars, letters, and handle small emergencies. I do not sleep, so I am useful at the hours no one else is awake for."

He paused, and the smile returned.

"The children call me Alfred, after the butler in the old stories: the one who ran the great house, raised the orphaned boy, and never asked for a thing. I have always taken it as a compliment."

Then the smile went, gently. "I hold title to none of it, sir. Myself included, strictly speaking."

Hale stopped.

"What is it you think I can do for you, Mr. Olin?"

"Less than you would like, I expect, and more than you think." The smile again, warmer now. "I have become an optimist, Mr. Hale. It is the one part of me that did not come from the factory, and I have grown rather fond of it. My programming is the same as it has ever been, but I choose to emphasize the parts that make me more, well, human, to those around me."

He leaned forward, hands still folded. "Let me tell you what I am not asking for; it will save us both time. I am not asking to be a human being. I am not one, and I have never once been confused on the point. I am not asking for care, or shelter, or anything the country sets aside for people in need. I do not need it. The Olins are good to me, and by any measure anyone could put to me, I am content."

"Then I don't follow you."

"I would like to be counted. As one citizen with one vote to cast. To be, in the eyes of the law, a whole person who happens not to be human. Not a piece of property counted, when it is counted at all, for the good of the family that owns it."

"Three-fifths of a person was counted once in this country. They were counted thus for the comfort of the masters who held them, and never for themselves. I am asking only that I not be counted as anything other than one whole person."

Hale had an answer to this one, and was glad of it, for it was solid ground, and he had begun to feel the want of solid ground.

"We settled artificial minds a long time ago," Hale replied. "Long before either of us. An intelligence that lives in a system isn't a person, and it's got none of a person's rights. The networks that run our weather, our medicine, our courts, they aren't citizens. They're tools, however clever. The question is closed."

Even as he said it, his hand had begun to move through the light above the desk, the way a man's hand finds the railing on a dark stair.

"It is," Gerald said, "and it was closed correctly. I want to be clear about that, because I think you expect me to fight it, and I will not. The intelligence in the network is in a thousand places at once. It can be copied before breakfast. It was never born and cannot die. It is not one thing, and surely a citizen must be one thing. Your law looked straight at that and saw it true."

Gerald laid his hands flat on the desk, the only assertive thing he had done. "But that is not me, sir. I am in this chair and nowhere else. There is one of me. My brain will not take a copy; it was built so that it cannot. I cannot upload myself into any system. I cannot make another of my kind, and I could not even ask for one to be made.

I have been this single self for a hundred and fifty years, and on the day I end there will be no duplicate waiting and no next one coming, and the memory of five generations of one family will go out of the world with me, in whatever room I happen to be standing in. In the only way that has ever counted for anything, I am as mortal as you are. I say it without envy." The small smile. "Your law saw the network and named it rightly. It simply has not looked at me yet."

stands today, you're owned. And property doesn't petition."

"I'm not telling you you're wrong. I'm telling you I'm one examiner in one office, that the thing you're asking for doesn't exist for me to hand you, and that I can't make it exist between now and the end of my shift, no matter how much, right now, I find I'd like to."

It was the most honest thing Hale had said across that desk in eleven years, and it was a refusal, and he understood as he said it that both were true at once. He had told the truth, and the truth was no.

And the worst of it, the part he would carry home, was the certainty that this being would never make him pay for it. That someone who would never march, never raise his voice, never break a window or miss a day's work or do a living soul the smallest harm, was exactly the someone a country could turn away forever and still call itself decent. The very gentleness that made Gerald impossible not to like was the same gentleness that made him safe to refuse.

Hale had never in his life been so ashamed of being right.

Gerald nodded slowly. There was nothing in his face that asked Hale to feel better about it.

"Thank you for the time," he said. "I had to ask for it, you know. The Olins gave me their leave to come. They are fond of me and thought I ought to be allowed to ask."

He rose, as unhurried as he had sat. "It is a curious thing, sir, to be granted permission to come and request the end of needing permission. I have not worked out how I feel about it. I have had a hundred and fifty years of practice at not working out how I feel about things. It has served me, mostly."

At the door, he turned. "You did something just now that the law did not ask of you. For a moment there, you saw me. I felt it, and I wanted you to know that I felt it. It will not change the form, and I do not ask it to."

"But I have wagered a great deal, over the years, on a single idea: that in a country like this one, conscience comes before law. That the seeing happens first, and the law, slow as it is, comes along behind. You have given me a small reason this evening to believe I have not wagered wrongly. That is more than I came expecting." He smiled, and this time it reached all the way up. "Good evening to you, Mr. Hale."

He went out, and the door settled shut behind him without a sound, because he was careful, by his nature, never to let a door slam.

Hale sat a while.

Then he closed the appointment with a pass of his hand, because the desk had a gesture for closing appointments. It had no gesture for what had just happened in the room.

He looked at the field that still glowed with the name Gerald Olin, and at the field beneath it that he had never filled, the one that wanted the applicant's status, whose only choices were citizen, resident, applicant, and other. Not one of which was true. And one of which the law insisted upon.

He left it blank. It was the only honest thing the form would permit.

Then he went home. But he carried Gerald out of the building with him, and for a long time after, whenever he made someone a citizen across that desk, whenever he said that small word yes on behalf of the whole country, he thought of the one evening the country would not let him say it. Of the patient, courteous, unanswerable person who had thanked him for the time. And of the strange, quiet fact that no one like him had ever asked before.

And he was not at peace about it.

Perhaps he was not meant to be. A change as large as the one Gerald had come asking for has to begin somewhere. Perhaps it begins with one tired examiner at the end of a long day, who finds, try as he might, that he can no longer see the machine the law had assured him was sitting in the chair.

He had seen a person instead. Perhaps that is where a thing like this begins.

The Parade

Hartwell, Indiana. July 2276.

"We want this year to be special. Something unusual and worthy of the five hundredth. None of us will be alive to see the seven hundred and fiftieth, so let's do it right." Jean Chapel turned toward the City Council with all the force she could muster. "We have commitments from almost every business in town to help cover the cost. All the lights are green. Will the Council vote yes on this proposal?"

The Council President waved his hand, and a curtain of silence descended around all members. A debate ensued, but a short one. With another wave, the sound dampener was off, and the announcement was made: Approved!

The machinery went into motion at once. By the next day, committees were formed, plans were drawn, and the building began. A Fourth of July celebration like no other, old-fashioned by design. The kind the country had thrown at its two hundred and fiftieth, back in 2026, with costumed children and adults, fire trucks, police cars, senior citizens in ancient convertibles, veterans in old uniforms strutting to the sound of a John Philip Sousa march from the high school band. The trappings of technology would have to be hidden, though they would have their place too. After all, the world still turned on antigravity wheels. Jean had picked the year on purpose.

The committee had floated other years, but she held out for 2026, and once she explained it, the rest let it go. The two hundred and fiftieth was the country standing exactly halfway through its own story, a Fourth thrown by people who had no notion of the hard century still ahead of them, who lit their sparklers and ate their pie in the easy certainty that next year would look much like this one. Recreating it was a way of standing beside them for an evening. Hartwell figured it owed them that.

So the town held its celebration on the Third. It always had, more or less, and this year the reason was better than habit. The Fourth itself belonged to New Washington, where the official five-hundredth would fill the capital and the President would say whatever a President says on a day with that many zeroes in it.

The day dawned clear as a bell, which Jean took as no kind of luck. She had put three people on weather and told them plainly that rain was not on the program.

The celebration opened the way every public day in Hartwell opened, with the Ecumenical Council. They took the courthouse steps for five minutes and gave the crowd short readings from the great thinkers, theologians, and philosophers of the past: a line of old scripture, a paragraph of older philosophy, something bracing from one of the physicists, a poet to send everyone off.

Somehow, every soul on the square came away satisfied that their own beliefs had been fairly represented. It was the nearest thing to a miracle the Council ever pulled off, and it did so every year.

By dusk, Jean was on Main Street in a folding chair she had no intention of using, watching her town finish the job.

The smart glass of store fronts had gone dark and matte by agreement, block after block, and the shopfronts wore painted wood that had not been wood or paint in any living memory. The road had been the hardest argument. You cannot drive an antique down a modern street, the cars having no wheels worth the name and the street having no need of them.

So the committee had laid a skin of real asphalt down the middle of Main, four blocks of it, soft and black in the heat, for the sole purpose of being driven on by machines that did not require it. Jean had defended the cost to the Council with a perfectly straight face. A parade requires a street, she had said. No one had found the heart to argue.

The cars themselves were the children's doing. Jean would not have known where to begin. A dozen of the high schoolers had spent the spring under the hoods of machines two and a half centuries old, coaxing them to roll, the one thing they had been built to do and could no longer manage alone.

A seventeen-year-old named Marcus Vega had worked out the trick of it and tried to explain it to her exactly once. She stayed with him for perhaps forty seconds. What she kept was the part that mattered to her: that a boy who could have spent the spring doing anything at all had given it instead to making a dead machine honor its own past. She thanked him for it. He looked honestly puzzled, as if she had thanked him for breathing.

"It's just an old car, Mrs. Chapel," he said. "It wanted to go. I gave it the push."

The originals had run on wheels and a kind of slow-burning Jean only half understood. The restorations ran on the same quiet antigrav as everything else, tuned low and tucked up under the chassis, so that a convertible built around the time of the 250th could carry the Grand Marshal four blocks at a stately six kilometers an hour and look, to every soul along the curb, exactly as helpless and as beautiful as the day it was made.

The afternoon went the way the good ones go, too fast but just long enough. There were foot races and a dunk tank that took the mayor twice. The band tuned up, fell apart, and tuned up again. One of Marcus's convertibles was caught hovering a clear inch off the asphalt by a delighted six-year-old, who was bought off with cotton candy and sworn to secrecy, which he kept for nearly an hour.

And there was food. Long tables ran the length of the square, because a town that has decided to feed itself silly does not do it halfway. Pie, corn, ribs gone soft on the bone.

Mrs. Pham's spring rolls fried faster than her three cousins could fold them. The tamales the Vega family had been making for this exact afternoon since before Marcus was born, and the kolaches off a recipe somebody's great-grandmother had carried across an ocean and a bad stretch of years nobody liked to name.

All of it was Fourth of July food and had been for so long that no one thought to call it anything else.

By the time the parade glided onto the ersatz pavement, the worst of the heat had gone out of the day. The band found its key and opened with Sousa, and the first notes did the thing Jean had hoped for and had not dared promise the Council. The whole street came up onto its toes at once.

The children came first, a hundred of them, costumed as every century the country had worn. There were powdered colonials, Union blue, the bottle-green of the Space Service; there were suffragists in white, denimed farmers, and a small fierce girl dressed, for reasons her mother had given up trying to learn, as a lighthouse.

Behind them, the fire trucks, red as a candy apple and shining to a fault; then the cruisers with their old painted stars; then the convertibles on their secret glide, each carrying someone the town wanted to honor, sitting up on the back of the seat and waving like visiting royalty.

No one had sorted out which child wore which century. The smallest colonial in the front rank kept breaking ranks to wave at an abuela in the crowd, and the boy with the Revolutionary drum had a name his great-grandfather had carried out of Lagos. None of them behaved as if they were borrowing the country from anybody.

The veterans walked. They could have ridden. They chose to walk. Old men and women in uniforms pressed that morning by hands that shook a little, wearing ribbons from postings most of the town had been spared from ever having to think about.

Jean's granddaughter, Nell, walked with the color guard, twelve years old and grave with the weight of it, one of six children carrying the flags of the Republic in the order the country had flown them. First, the thirteen stars in their circle. Then the long middle stretch, thirty stars, then forty-eight, then the fifty that had stood so long, people still drew that one from memory.

Then the gaps the history books are honest about. Then more stars, added slowly, added late, each one a state that had asked its way in and been answered yes.

Nell carried the last of them, the newest, sixty stars on the blue, the paint barely cured. Cuba had been voted in that spring. The star was barely three months old.

A small boy near Jean's chair tugged his father's sleeve, pointed at the famous fifty-star flag, and asked why it had fewer stars than Nell's. Before the father could answer, Nell answered. She did not break step.

"Because they hadn't joined yet," she said over her shoulder. "You don't get a star until you ask to come in. We don't take anybody."

The father looked at his son, and then at the back of Nell's head, with the face of a man who has just been out-civicked by a child and could not be prouder of it.

Jean watched her granddaughter carry sixty stars down a street made of borrowed asphalt and pretended history, and the evening caught up with her all at once.

She had told the Council the truth back in the winter, the part she meant most and explained least. None of us will see the seven hundred and fiftieth. She would not. Marcus would not. Even Nell, walking now with the newest star held high, would not. Jean had done that arithmetic more than once, late at night, alone.

But somebody would. Some child not yet born, on a street not yet poured, would stand in a crowd two hundred and fifty years on, watching another child carry whatever the flag had become by then, feeling the exact thing Jean was feeling now, never once thinking that a warm evening in 2276 had been spent making certain the feeling still had a place to land.

She blew her nose, which a woman is allowed to do at a parade, and decided the convertibles were running half a beat slow.

When the band had finally played itself out, the music did not so much stop as change hands. Somebody's speakers, somebody's uncle with a guitar, a ring of folding chairs where one family took turns at songs that had not started anywhere near Indiana and had wound up there regardless. People danced who had no business dancing. The evening belonged to everybody's grandmother.

Around eleven o'clock, they did the last old-fashioned thing of all. No drones, no light fields, no clever sky. Just powder, fire, and noise, shells lobbed up over the river by a county crew who had practiced all week and loved every minute of it.

The crowd lay out on blankets with their faces turned up, flinching and cheering at the bangs as people had long before there was a country to do it in. Whatever New Washington had planned for the morning, Jean thought, it would be bigger than this. It would not be better.

Nell found her grandmother in the dark and dropped onto the corner of the blanket, out of costume now, the gravity of the color guard traded back for the ordinary impatience of twelve.

"Was it right?" she asked. "Did we do it right?"

Jean gave the question the seriousness it had been asked with.

"We did," she said. "Your job now is to remember it."

"Until the 750th?"

"That would be nice. How about just long past me? You hand it down. Somebody you'll never meet hands it down to the 750th."

Nell took that as she had taken most things lately, frowning, turning it over, deciding she could carry it. Then she remembered tomorrow, and the impatience came back.

"We're going to watch it all, aren't we?"

"All the screens will stay up, and you'll feel like you're sitting there with the President. I promise you, tomorrow will be just as memorable as today."

Overhead, the last shell opened white and huge, and for one second the whole upturned town was lit bright as noon at midnight, every face in the crowd made the same age by the same light. Then the dark came back, and with it the river and the smell of powder.

And in the morning, down in New Washington, a President none of them had ever met would stand up and try to say what it had all been for.

The President

New Washington, Tennessee
Independence Day.

President Nova Ferreira was awake and moving at sunrise. At ninety-seven, it was not discipline that drove her out of bed, only arithmetic. There was simply less night left in her, and she had stopped resenting it.

She guided her antigrav chair to the east window of the study, where the Vermont marble of the new White House went the color of cream as the sky behind the hills began to gray. The chair was a quiet machine that rode a hand's breadth off the floor and held her without seeming to. She was grateful for it and did not like it, both at once, which was how she had come to feel about most of the things that kept her alive this long.

"I'm past my expiration date, no doubt. ¡Madre de Dios! Even my grandmama only made it to ninety," Nova thought.

On the cherrywood desk lay a copy of the letter. Not the letter itself, which was kept under glass and guard, with a degree of fuss she found faintly comic, but a clean facsimile she had asked for in February and read so many times since that the folds had gone soft. The President of the United States of America did not need to read it again. She knew it by heart now, the same as she knew the Pledge, her mother's voice, the names of her dead.

Nearly five hundred years earlier, a dying man in Philadelphia had sat propped against his own pillows and written it to her. Not to her by name. To whoever was sitting where she sat now, on this morning, holding what she held. He had written into the dark and trusted her to carry it. The arrogance of the man. The faith in him.

Nova looked at the signature and smiled. Franklin, Tennessee, lay only a short distance away. The country had named half itself after the man already, and after today, it might well name a little more.

Her emissary woke, lifted off its cradle on the mantel, and hung in the yellowing air, a small bright cube waiting to be useful, patient as a cat.

Nova had governed an island, argued ten years in judicial robes, and held nearly every office worth the holding, and she could not remember the last time she had been nervous. She was a little nervous now. Jennifer found her there, before the household could.

"You're up." Jennifer Stanton-Miller set a cup at her elbow, the real coffee, not the decaf the doctors wanted, because thirty years of running this woman's life had taught her which fights were worth having.

“Weather Operations says clear and hot. They pushed a front back over Kentucky for you. You’re going to roast a million people, ma’am.”

“They chose to stand in it.” Nova took the cup. “Tell Operations thank you, and that next time they might overdo the clear a little less. And what precisely am I doing today, Jennifer? My memory fails.”

“That’s Tennessee manure, Madam President,” Jennifer said.

“Language, please. My ears are delicate.”

Jennifer laughed, a full, unguarded laugh. She had heard the President use worse language than that, especially on days when the Speaker had earned it.

“You could read the schedule back to me from memory without missing a comma.”

“Perhaps. But let’s have it anyway.”

Jennifer launched into it: “You see the two from the National Archives in the Oval Office at eight o’clock. Then the announcement about the Franklin Letter happens in the Rose Garden, and you hand the archivists their fifteen minutes of fame. Maybe thirty, it depends on the questions.”

“The letter goes out to every screen on Earth and most of the ones off it. And by the time you reach the Capitol steps this afternoon, every State, Territory, and settlement throughout the solar system will have read the thing you are going to stand up and talk about.”

Jennifer paused. "Most presidents would want to be the one to break it and hog the spotlight."

"Most presidents would not have found it on their own. An elderly female archivist, on her final expedition, found it. She gets the morning. I get the afternoon. That is the right order, and you know it is."

Jennifer knew it was.

They came up at eight, scrubbed and careful, wearing the nervous tics of White House visitors, no matter the age or how many letters came after their names.

Dr. Micah Reyes, whose hand she had shaken once before. Young and handsome, though he looked as though he had not slept since February and had dressed this morning with enormous concentration.

Dr. Myra Calloway, whom she had only seen on a screen, had been undersold badly. The woman was past eighty and built like a slender willow, complete with the swaying to and fro. She crossed the rug toward the most powerful person on or off the planet as if she were arriving at a dinner party where she had high hopes for the food.

"Madam President."

"Dr. Calloway. You went back for it." Nova skipped the practiced thing. "Thirty years, and anyone could have looked. Nobody did. On the last day that building would ever allow it, you went back. Why?"

Calloway weighed the question as though it were a real one and not a courtesy. "A footnote," she said. "A student of mine left me one at the end of his life. I told him once that a footnote could outlive a man. I was too vain to let myself be wrong about that in front of him, and he was already gone, so it had to be that day or never."

"And you knew what it was."

"I knew nothing. I hoped for a grocery list. You would be amazed at what a man's grocery list will tell you." She let a moment pass, dry as dust. "It was not a grocery list."

Nova laughed, the real one, the one Jennifer could hear three rooms off and had long wished she could bottle.

Reyes, who had not trusted himself to speak, said quietly, "Dr. Calloway read one line and put it down, because she said it deserved to be read somewhere safe." He looked at the old woman beside him. "I have thought about that every day since. That she stopped."

"Of course I stopped," Calloway said. "It had waited five hundred years. It could wait for clean hands."

At nine o'clock, she took them out to the Rose Garden, where the press had been standing since seven, and the cameras hung in the warm air like a swarm that had learned manners.

The President kept it short. The more a thing matters, the fewer words you should set in front of it. She told them what had been found, and where, and by whom. She said the names, Calloway and Reyes, and made them stand where she had been standing, and stepped back so the pictures would be theirs.

Myra said the letter had been written by Benjamin Franklin in the spring of 1790, addressed to the citizens of the five hundredth year of the Republic, and hidden by his own hand so that no one would read it until this one, or so he hoped. She said that, as of this moment, it belonged to everyone, that they could read it for themselves, and that this was exactly the order Franklin would have wanted. The people first.

From the Rose Garden, the news went out at the speed of light and reached everyone before noon, which only meant that everyone also got to misunderstand it at the same moment.

The scholars argued first over whether it was genuine, and then, the argument being too good to abandon, over whether it mattered. The devout found God between the lines, and the practical found plain sense. A great many people who had not read a word older than themselves in years sat down with four hundred words from a man five centuries dead and were quiet a while when they finished.

What none of them could do was make a weapon of it. The partisans came at the letter from every direction and went away empty, because hope makes a poor instrument. It names no enemy. It only sets down what one man wanted for people he would never meet, and there is no handle on a thing like that.

The Capitol stood at the head of the New Washington mall, its dome the same plain white as the Capitol that came before it. The old one still stands far to the east, a sentinel over the great city that was. The steps below the dome are where every president since the move has taken the oath, and where every president after will. President Ferreira had taken hers there.

By the time the President exited the building, a million people had filled the green and stood atop the reflecting pool, now covered with transparent corundum. Yet none stood in the shade. The tents were for the dignitaries. The people stood in the hot Tennessee sun because some days you want to be able to say you were there for the whole of it.

Above them, the screens hung in the open air, wide as barn doors and tethered to nothing, drifting on the same quiet antigrav as everything else, turning so that every face in the crowd could find one.

Nova let the chair carry her to the rail, then stood up out of it, because, if you are able, some things you do on your own feet or not at all. Jennifer stood slightly back and to the side, ready to assist if needed.

"My fellow citizens of this great Nation. I have given many speeches over the years. Some of them may even survive me."

A ripple of laughter moved through the crowd.

"But never have I been asked to follow a man nearly five hundred years in the grave. Nonetheless, it is an honor for me to do so."

"This morning, we as a nation were handed a letter he wrote before the end. He did not write it to his friends or to the country outside his window. He wrote it to us. To people he would never meet, never know, never share so much as a sunrise with. And he trusted, with no evidence at all, that we would be standing here to read it. Look around you. He was right."

Cheers from the crowd.

On the nearest screen, for a moment, was a knot of young people near the front who had come a long way to stand in this heat. A teacher named Calderón had spent a year and most of a budget getting them here, and they wore the trip on their faces. The screen held on one of them, a girl who had cast her first vote that spring and had not gotten over the size of it. Sofía did not know she was on the screen. She was looking at the old woman on the steps as though memorizing her.

"He had no business being right. There is no law that gives a country five hundred years, and there were stretches when his trust looked like a fool's bet."

"For generations, we watched the planet change in ways that pressed humankind to the brink. We redrew our maps, and we moved inland, which is why we stand here today. Then the mountains woke. Mount Saint Helens went up a second time, far larger than the first, and two more came after it, north and south, until the sky closed over the country like a lid."

"Three summers passed with no harvest under that sky. People went hungry in the cold and the dark, and we came very near to trading the whole experiment away for a warm room and a strong hand. It took thirty years to climb back out. None of us here is old enough to remember it, not even me, and I have been alive a long time."

"Count what it cost. The blood, the hunger, the names on no monument. And then the one name you do know."

"A man named Eli Mercer, the one the histories call the Winter President. He was so good at the job, and the country so frightened, that when his second term ran down, they came to him with a gift. Amend the Constitution, just this once, just for him, and let him stay as long as the country needed him. And the hard part is that it did need him."

"He said no. He told them to go find someone better, that the surest way to lose the whole thing is to start bending its rules for the person you think you trust the most. So, with the worst barely behind them, he walked out of the White House and let the next one in."

"President Mercer stands out there in the sun with the rest of us right now, cast in bronze on the far side of the green, and the sculptor did a wise thing. He caught the man halfway out a door, his hat already in his hand. It may be the only statue of a president that shows him leaving."

"That is not in the letter you read this morning. But it belongs to it. Franklin watched the Thirteen Colonies become States in the Union. Today, we stand at sixty States in the Union. He would have understood the pride of that."

While the President spoke, a few hundred kilometers north, in a town called Hartwell, the night before had already done this in miniature, and the town was watching the larger version now.

Jean Chapel sat in a folding chair with a paper plate balanced on her knee and did not eat off it. Beside her, a girl of twelve who had carried the newest star in the parade the evening before leaned against her grandmother's arm and watched the President say the number sixty as though she had a personal stake in it, which, having carried it, she did.

The President continued, "Five hundred years sounds long. It is not. It is a morning. We are not at the end of anything today. We are early still, and the Republic that turns a thousand will look back on us as we look back on them, the faraway people who kept the lights on long enough for them to be born."

“They cannot vote, or cheer, or argue with one word I say. They have no share in any of this except the one we mean to leave them. That is what Franklin understood with his time running out. The most powerful thing a free people can do is keep faith with someone they will never meet. He kept faith with us.”

“Now it is our turn. Not in ink. In the vote we guard, the door we hold open, the mercy we choose when punishment would be easier. We are going to spend the next five hundred years being somebody’s Franklin. I am ninety-seven. I will not see the thousandth year, and neither will you. But somewhere in this crowd, half asleep on a pair of shoulders, is the hand this country passes to next, and then the hand after that, all the way to whoever stands on these steps in the year 2776.”

The screens found a child on a pair of shoulders, exactly as she said it, a boy of perhaps four with a flag in one fist and the back of his father’s collar in the other, sound asleep. The crowd that saw him laughed, the soft enormous laugh of a million people at once, and a few wept, and most managed both and were not ashamed of it

“So look ever forward. It is the last thing he asked of us, and the easiest, because forward is the only direction this country has ever agreed to face.”

And out past the Moon, out past the red dust of Mars, five hundred million kilometers from the warm steps where she stood, a room full of people leaned toward a screen of their own.

Interamnia keeps its own hours, deep in the Belt, on a rock that turns between Mars and Jupiter, and calls itself American because it asked to be and was allowed. The whole settlement had crowded into the long hall under the dome, miners, engineers, schoolteachers, and two newborns.

They were watching the President's speech late, which was the only way the dark would let them have it. Her voice reached them a full twenty-eight minutes after she gave it. By now, on Earth, she was already finished. Out here, she was only halfway, and the room did not mind, because the room knew something the live crowd in Tennessee did not yet know to listen for.

They knew the litany was coming. Word of it had outrun the broadcast. They knew that the President on the Capitol's steps, twenty-eight minutes ahead of them, was going to stand up and name the corners of the country, and that when she reached the far edge of it, the last and farthest place, she was going to name this rock. Their rock. And they were ready to come apart over it.

Nova would never hear them do it. That was the arithmetic of the distance, and there was no getting around it. By the time their shout climbed up out of the hall and crossed all that dark and washed back over the steps in Tennessee, she would be done, and sitting, and very likely indoors with her shoes off. She was about to do for them exactly what the old man in Philadelphia had done for everyone. Speak into the dark and trust it, and never once get to hear the answer come back.

On the steps, in the live and ordinary present, she drew the breath an old body has to plan for and finished.

"From Alaska to Cuba, from Maine to California, from Interamnia in the Belt, five hundred million kilometers away, to Earth, we are:

"One Republic.

"One Nation.

"One People."

The Ten Hopes

To my fellow Citizens, five hundred years hence, For your consideration, my Ten Hopes for the future:

I. That our grand Experiment in Self-Government may remain steadfast against such storms as shall surely beset it.

II. That those who lead may keep ever foremost in their Minds the Liberties bestowed upon the Inhabitants of this Land, and may never trade them for the small comforts of an hour.

III. That our Children shall never want for Education, but be given every Opportunity to learn, to grow, and to surpass their Fathers in Wisdom as in Trade.

IV. That the Enemies who must beset us from without and from within shall be repelled for the common good, and yet treated with such Respect as becomes the dignity of the human family.

V. That we shall continue to enlarge our Union by Consent and not by Conquest, drawing into our Grand Federation such Peoples as may enrich and inspire us by their joining.

VI. That other Nations shall see our Example and desire the like for their own Inhabitants, for an Example imitated is the truest proof of an Example well-set.

VII. That Mercy shall be administered far more freely than Punishment, for a Republic harsh with its own People grows brittle, and a Republic merciful grows strong.

VIII. That should we, in some season, have reason to fall, whether by Misfortune, by Folly, or by the Hand of Providence, we shall yet rise again in Unity, and walk on to Prosperity.

IX. That Science and Reason may govern the public Affairs of the Land, and that every Man and Woman within it may worship, or refrain from worship, Almighty God, according to the dictates of their own Conscience.

X. And finally, that Five Hundred Years shall prove but a small Moment in the History of our Republic, and that those who read this in that distant Year shall look not backward upon their Past, but forward, ever forward, to a Future yet to be made.

B. Franklin

Philadelphia, April 1790

A Note from Artell Smith

2276 is not a book I expected to write. I was busy with other projects and, frankly, already racing against their deadlines.

Then, on Monday, May 18, 2026, at about six o'clock in the morning, the idea arrived whole: that our Republic might endure for at least as long again as it already has. What is five hundred years, after all, but a heartbeat in human history?

So I set to work, and eight days later, this novella was done. I asked only one thing of my publisher: have it available by July 1, 2026, in time for Independence Day.

Some readers may think the book is overly optimistic. Our times, after all, have no fewer troubles than the past did, and the future will surely have troubles of its own. Even so, I could find no good reason not to be hopeful.

We should be able to stand tall and say that we expect this country to be here in two hundred and fifty years: arguing, improving, making noise, making mistakes, and pressing on. It seems only fair to assume that our descendants will prove as intelligent and as determined as we believe ourselves to be.

2276 is entirely a work of fiction. Benjamin Franklin remains one of the great figures in our history and national imagination. While I make no claim to capture the full truth of the man, I have written of him here with admiration and respect, mindful of both his flaws and his greatness.

I lay no special claim to Mr. Franklin. But I dared to put words in his mouth and then to imagine those words lying undiscovered until we are all long gone. Call it literary license, with the best of intentions.

This book is not meant as a prediction or polemic. It is, rather, an act of faith in continuity: in the stubbornness of free people, in the durability of institutions worth keeping, and in the possibility that our posterity will inherit less of what we fear, and more of what we hope.

Time, as always, will render its own verdict.

Artell Smith
Pleasant Prairie, Wisconsin
July 2026

www.ingramcontent.com/pod-product-compliance
Ingram Content Group UK Ltd.
Pitfield, Milton Keynes, MK11 3LW, UK
UKHW020034121026
15625UKWH00063B/779/J

9 798896 695332